WITHDRAWN

JOURNEY INTO SPACE

The Inner Planets

by Christina Leaf
Illustrated by Natalya Karpova

BLASTOFF! MISSIONS

BELLWETHER MEDIA
MINNEAPOLIS, MN

Blastoff! Missions takes you on a learning adventure! Colorful illustrations and exciting narratives highlight cool facts about our world and beyond. Read the mission goals and follow the narrative to gain knowledge, build reading skills, and have fun!

Traditional Nonfiction

Narrative Nonfiction

Blastoff! Universe

MISSION GOALS

> FIND YOUR SIGHT WORDS IN THE BOOK.

> LEARN THE ORDER, NAMES, AND FEATURES OF THE INNER PLANETS.

> THINK OF QUESTIONS TO ASK WHILE YOU READ.

This edition first published in 2023 by Bellwether Media, Inc.

No part of this publication may be reproduced in whole or in part without written permission of the publisher. For information regarding permission, write to Bellwether Media, Inc., Attention: Permissions Department, 6012 Blue Circle Drive, Minnetonka, MN 55343.

Library of Congress Cataloging-in-Publication Data

Names: Leaf, Christina, author.
Title: The inner planets / by Christina Leaf.
Description: Minneapolis, MN : Bellwether Media, 2023. | Series: Blastoff! missions. Journey into space | Includes bibliographical references and index. | Audience: Ages 5-8 | Audience: Grades 2-3 | Summary: "Vibrant illustrations accompany information about the inner planets of the solar system. The narrative nonfiction text is intended for students in kindergarten through third grade"-- Provided by publisher.
Identifiers: LCCN 2022006861 (print) | LCCN 2022006862 (ebook) | ISBN 9781644876558 (library binding) | ISBN 9781648348396 (paperback) | ISBN 9781648347016 (ebook)
Subjects: LCSH: Inner planets--Juvenile literature.
Classification: LCC QB606 .L43 2023 (print) | LCC QB606 (ebook) | DDC 523.4--dc23/eng20220422
LC record available at https://lccn.loc.gov/2022006861
LC ebook record available at https://lccn.loc.gov/2022006862

Text copyright © 2023 by Bellwether Media, Inc. BLASTOFF! MISSIONS and associated logos are trademarks and/or registered trademarks of Bellwether Media, Inc.

Editor: Besty Rathburn   Designer: Jeffrey Kollock

Printed in the United States of America, North Mankato, MN.

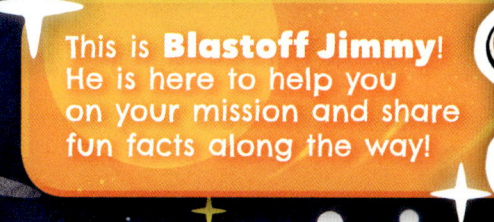

This is **Blastoff Jimmy**! He is here to help you on your mission and share fun facts along the way!

Table of Contents

Above Earth 4
Mercury 8
Venus 12
Mars 16
Glossary 22
To Learn More 23
Beyond the Mission 24
Index 24

Above Earth

Your plane flies high above Earth. Through the clouds, you see a patchwork of green land and blue water below.

Earth is a rocky planet, but it looks **lush** from here!

You think about how Earth's **atmosphere** is unique. It allows life to grow.

What makes the other planets between the Sun and the **asteroid belt** stand out? You head off on an imagination quest to the rocky inner planets!

Hmm...

Mercury

You put on sunglasses as you arrive at Mercury. The Sun is so close!

8

You are surprised at how small Mercury is. It is only a little bigger than Earth's moon!

Mercury

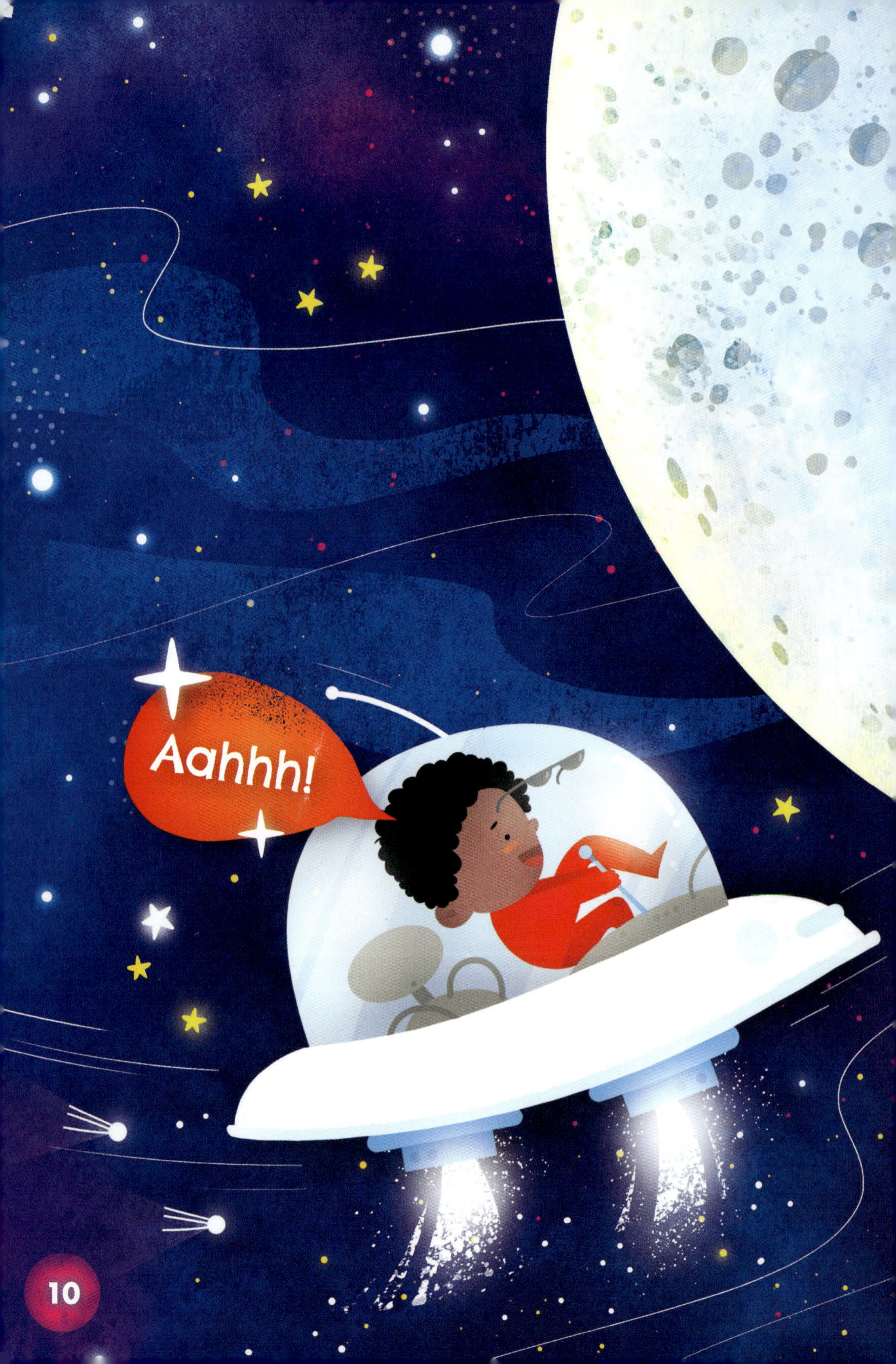
Aahhh!

craters

You speed up to keep pace with the **cratered** planet. Mercury **orbits** quickly! But soon you decide to move on. You do not want a sunburn!

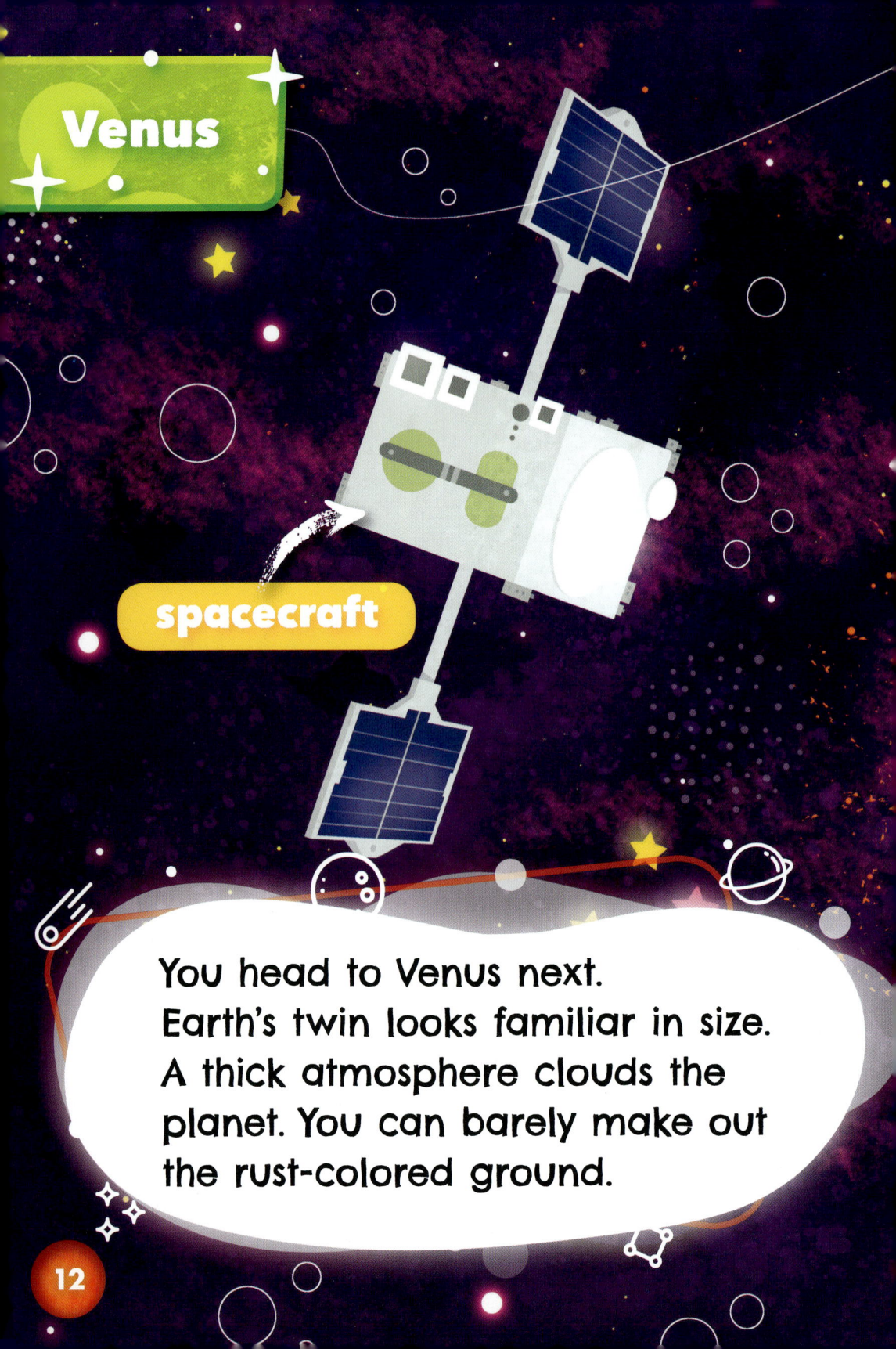

Venus

spacecraft

You head to Venus next. Earth's twin looks familiar in size. A thick atmosphere clouds the planet. You can barely make out the rust-colored ground.

JIMMY SAYS

Venus is the hottest planet. It reaches 900 degrees Fahrenheit (482 degrees Celsius)! Spacecraft do not last on the planet. They orbit it instead.

Venus

volcano

Finally, you spy **volcanoes** and mountains. But you start to cough as you get closer. Venus stinks! The smell seems like it might be **toxic**. Time to go!

Cough!

Mars

You wave at Earth on the way to Mars. Soon, the Red Planet comes into view. Two moons circle around it. An **ice cap** tops the planet.

Deimos

ice cap

Mars

Phobos

17

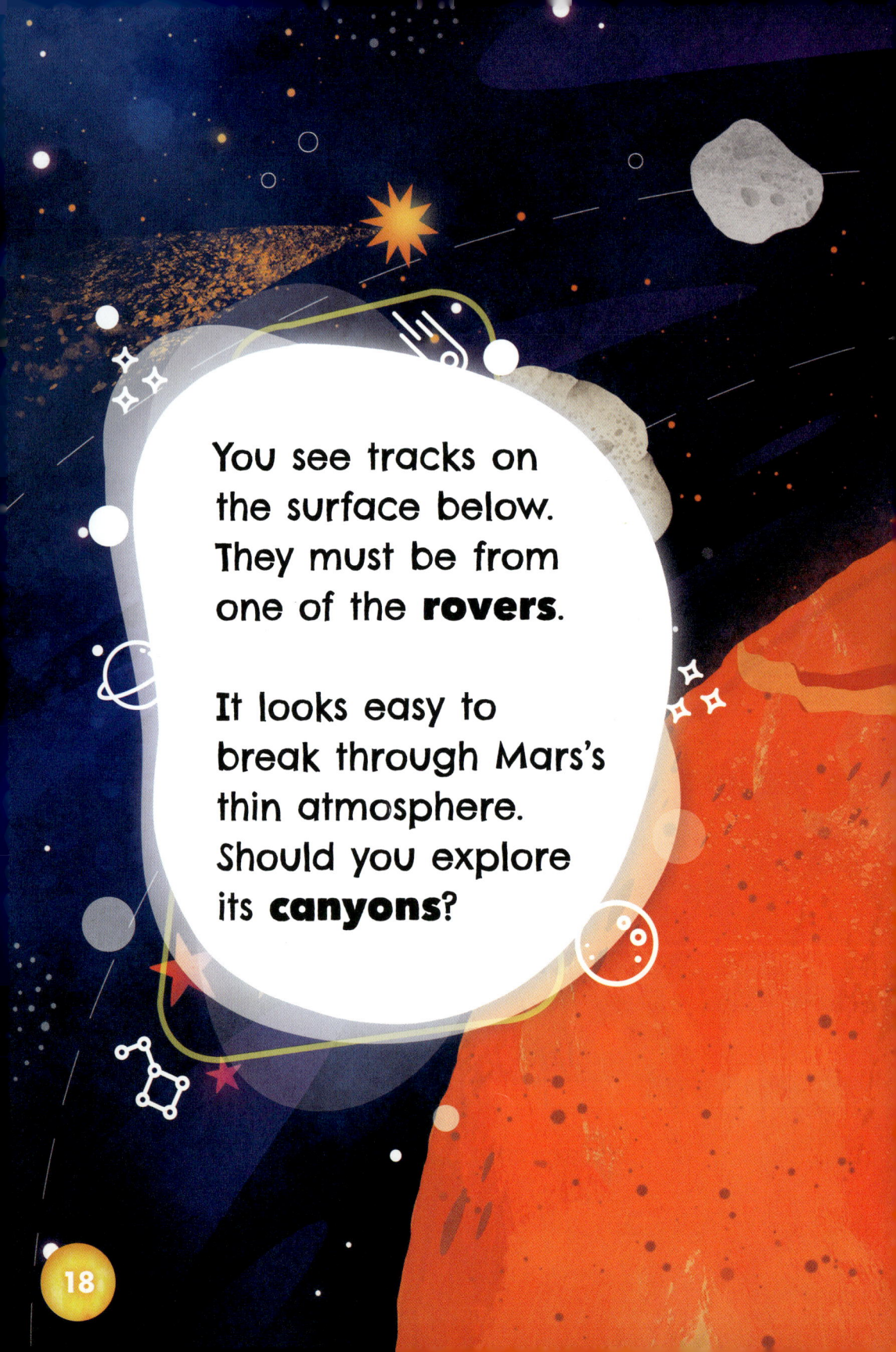

You see tracks on the surface below. They must be from one of the **rovers**.

It looks easy to break through Mars's thin atmosphere. Should you explore its **canyons**?

JIMMY SAYS

Six rovers have landed on Mars. The most recent from the United States is looking for signs of life!

rover

canyon

A bump jolts you out of your daydream. Your plane is landing. It was a good trip to the inner planets. But it feels great to be home on Earth!

The Inner Planets

Known for: smallest, closest to the Sun, fast orbit

Planet Size Rank: 8

Number from the Sun: 1

Mercury

Venus

Known for: thick atmosphere, similar size to Earth, volcanoes

Planet Size Rank: 6

Number from the Sun: 2

Earth

Known for: green land, blue waters, life

Planet Size Rank: 5

Number from the Sun: 3

Mars

Known for: Red Planet, canyons, rover exploration

Planet Size Rank: 7

Number from the Sun: 4

Glossary

asteroid belt–a part of the solar system between Mars and Jupiter where more than one million asteroids are found

atmosphere–the gases that surround a planet

canyons–deep, narrow valleys with steep sides

cratered–having holes in the surface

ice cap–a very large area of snow and ice that covers the top of a planet

lush–covered in healthy green plants

orbits–moves in a fixed path around something

rovers–vehicles that explore the surface of a planet or moon

toxic–poisonous or harmful

volcanoes–holes in the ground; when a volcano erupts, hot ash, gas, or melted rock called lava shoots out.

To Learn More

AT THE LIBRARY

Betts, Bruce. *Super Cool Space Facts: A Fun, Fact-filled Space Book for Kids.* Emeryville, Calif.: Rockridge Press, 2019.

Edson, Shauna, and Giles Sparrow. *Mars: Explore the Mysteries of the Red Planet.* New York, N.Y.: DK Publishing, 2020.

Leaf, Christina. *The Outer Planets.* Minneapolis, Minn.: Bellwether Media, 2023.

ON THE WEB

FACTSURFER

Factsurfer.com gives you a safe, fun way to find more information.

1. Go to www.factsurfer.com.

2. Enter "inner planets" into the search box and click 🔍.

3. Select your book cover to see a list of related content.

BEYOND THE MISSION

> WHICH INNER PLANET WOULD YOU LIKE TO VISIT? WHY?

> CREATE AN INNER PLANET. WHAT IS ITS NAME? WHERE IS IT LOCATED?

> MAKE A LIST OF ITEMS YOU WOULD BRING IF YOU TOOK A TRIP TO MARS.

Index

asteroid belt, 7
atmosphere, 6, 12, 18
canyons, 18, 19
craters, 11
Earth, 4, 5, 6, 9, 12, 16, 20
ice cap, 16, 17
land, 4
life, 6, 19
Mars, 16, 17, 18, 19
Mercury, 8, 9, 11
moon, 9, 16
mountains, 15
orbits, 11, 13
rocky planet, 5, 7

rovers, 18, 19
size, 9, 12
spacecraft, 12, 13
Sun, 7, 8
United States, 19
Venus, 12, 13, 15
volcanoes, 14, 15
water, 4